Stryker MGS

M1128 Mobile Gun System

Warmachines 02 is a visual reference of the U.S. Army Stryker Mobile Gun System in combat in Iraq and Afghanistan.

VP
[VE]RLINDEN PUBLICATIONS

SABOT Publications

Warmachines 02 Stryker MGS

SABOT Publications
941 Cottontail Lane
St. Charles, Missouri 63303
United States of America
ISBN: 978-0-9973774-7-7

Special thanks to our friends for their valuable contributions to this book:
Francois Verlinden - for his inspiration and dedication to the hobby
Bjørn Weber
LTC Kim Hartvig Sørensen
Robert Burik
Chun Hsu
Jeff DeRosa
Chuck Sherman
John Charvat
Mike Petty

Scouts, Tankers and Mechanics everywhere
and all of the Soldiers, Sailors, Marines and Airmen who serve
Thank you!

Chris Mrosko
Brett Avants
Pearce Browning
SABOT Publications LLC

Printed in the United States of America

Introduction

The M1128 Mobile Gun System (MGS) is a weapons platform based on the Stryker family of wheeled armored vehicles. It mounts a 105mm M68A2 rifled tank gun. The main weapon is fully stabilized and is fed by a fully automatic loading system. While the MGS is not an anti-tank platform, it can engage and destroy tanks and armored vehicles. It's primary role was initially to support the quick-reacting but lightly-armored airborne infantry units. It was a compromise between heavy firepower and lightweight nimbleness.

The first MGS units deployed to Iraq in 2008. The cannister rounds fired by the 105mm main gun were a big hit with troops. However, there was a lot of equipment packed into a tiny space. The MGS doesn't really have a full turret, but more of a remote weapons station with a crewman on each side of the gun. The vehicle carries 18 rounds of main gun ammo, along with 400 rounds of .50 cal and 400 rounds of 7.62mm for the two machineguns.

While the MGS has received a bad rap in some circles, others, especially those who operate the vehicle in combat, have different opinions. One senior NCO has stated, "I cannot stress enough that it is not a tank; it is a support vehicle with some tank-related features. This vehicle is fast, maneuverable, quiet and accurate. In my opinion, it is the most lethal ground vehicle for an urban environment in Iraq today." The thermal imaging system is a big advantage on night missions, and the cannister rounds are lethal to any group of terrorists that want to travel together. The slat armor also helps protect the vehicle from RPGs and other threats.

The MGS can be a lethal vehicle when deployed properly.

Stryker MGS

In Combat

·tryker Soldiers with Company B, 4-9 Infantry, walk through a village in the outskirts of Baqubah, as he Mobile Gun System (MGS) Stryker variant pulls security down a road. Soldiers of 4-9 Infantry, 4th ·tryker Brigade Combat Team are assisting Soldiers of 3rd Stryker Brigade Combat Team in the clearing ·f Baqubah, a major operation known as "Arrowhead Ripper," by isolating Baqubah just outside the ity limits, to prevent insurgents from getting in, or out, of the city.

ıldiers fire their weapons from a MGS Stryker variant on the outskirts of ıqubah, Iraq. Strykers have sealed off Baqubah and continues to clear the ty of Baqubah in a major offensive known as Arrowhead Ripper, launched ı the city and its outskirts. Arrowhead Ripper aims to kill or capture all surgents in Baqubah by sealing off the city and any avenues of escape.

First Brigade Combat Team "Bastogne", 101st Airborne Division (Air Assault) leadership talk about M1128 Stryker Mobile Gun System Vehicle capabilities with 4-9 Infantry Battalion, 4th Brigade Combat Team, 2nd Infantry Division Soldiers in Salah ah Din province.

A M1128 Stryker Mobile Gun System from 2nd Battalion, 1st Infantry Regiment, 5th Stryker Brigade Combat Team, 2nd Infantry Division convoys along highway 1 in Hutal, Kandahar province, Afghanistan, Dec. 25, 2009.

ABOVE and BELOW: U.S. Soldiers from 2nd Battalion, 1st Infantry Regiment, 5th Stryker Brigade Combat Team, 2nd Infantry Division test fire the M1128 Stryker Mobile Gun System in Hutal, Afghanistan, Jan. 21, 2010.

BAY
MG AMMO
AMMO

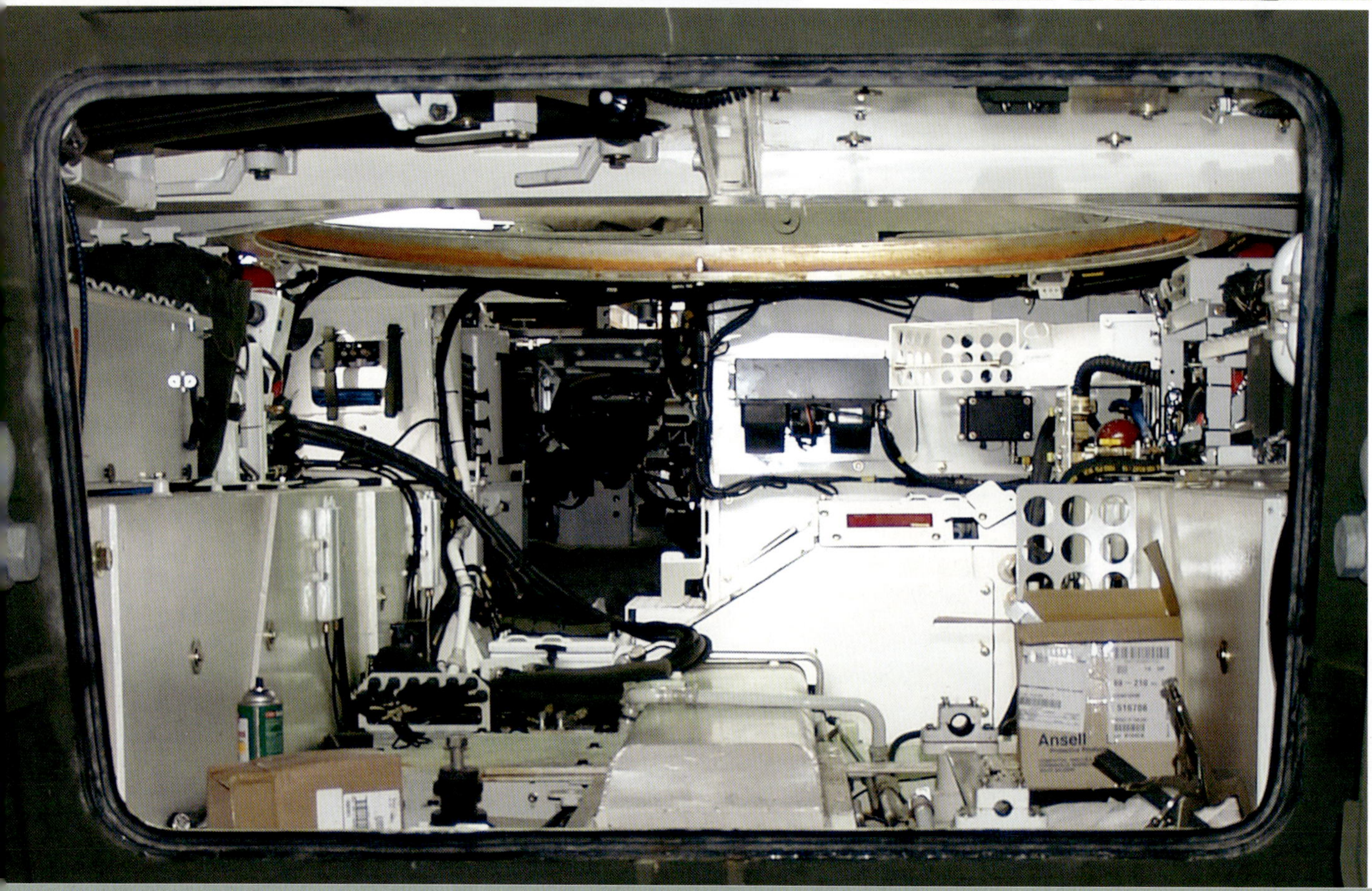
Ansell

Early MGS, Ft. Lewis, Washington, September 2007

PAGE 8 ABOVE: The Stryker Commander's Thermal Viewer in its folded position at the rear of the turret.

PAGE 8 BELOW: A shot of the MGS interior from the rear hatch, looking at the driver's compartment.

ABOVE: The MGS turret undergoing maintenance.

BELOW: The main gun ammunition carousel.

Stryker MGS

Walkaround

ABOVE and LEFT: A couple of good shots of the right front of the MGS with slat armor. The wheels on the Stryker family have been noted as a vulnerability, and this armor package address that issue with a number of armor sections covering the top half of the roadwheels.

The flag-looking device is a Rhino Convoy Protection Device. Rhino is a capability used to defeat passive infrared-triggered (PIR) improvised explosive devices (IEDs) and features a universal bracket that can be mounted to all combat and tactical vehicle platforms.

ABOVE: The mine roller is The Self-Protection Adaptive Roller Kit (SPARK), a modular mine roller system designed to be mounted on tactical wheeled platforms.

RIGHT: A close-up view of the slat armor and headlight assembly on the left front of the vehicle.

ABOVE: Another look at the armored side skirts covering the wheels, and the slat armor covering the entire chassis. The wheels on the Stryker family have been noted as a vulnerability, and this armor package address that issue with a number of armor sections covering the top half of the roadwheels. Note the layered armor sections.

LEFT: A rear view of the antenna assembly on the MGS, including the rear of the turret and the slat armor.

RIGHT and BELOW: Two good views of the rear of the MGS showing the slat armor configuration and the slat-armored side skirts. Slat armor was developed as an inexpensive and effective means of protecting the lightly armored MGS from rocket propelled grenades (RPGs) and other shaped-charged infantry weapons.

RIGHT: Smoke grenade launchers are mounted on the MGS on either side of the gun tube.

BELOW: A left front view of the slat armor mounting brackets.

RIGHT: The Rhino Convoy Protection Device. Rhino is a capability used to defeat passive infrared-triggered (PIR) improvised explosive devices (IEDs). Continuously fielded since 2007, more than 34,000 Rhinos have been delivered to theater.

BELOW: The Rhino mounted on the front of the MGS on the slat armor.

ABOVE and RIGHT: Several right side views of the MGS showing the slat armor configuration and other equipment commonly carried on the vehicle. Note the dirt and dust on the MGS, and the streaking on the sides.

ABOVE: The mount for the M2 .50 cal machine gun. The M2 can provide supporting fire for infantry when the main gun may be overkill.

LEFT: A close-up of the commanders panoramic viewer and the wind sensor.

The MGS provides a distinct advantage to its supporting infantry with its excellent night vision capabilities and its deadly accurate 105mm M68A2 main gun.

ABOVE: A close-up view of the slat armor. Again, note the intense weathering, with dirt and dust covering most of the green vehicle.

LEFT: A close-up view of the urban survival kit components showing the protective shields. This vehicle really needs a bath!

RIGHT: a rear view of the antenna mounts on the MGS. It's hard to believe that this is a green vehicle.

BELOW: A close view of the commanders station.

Pages 22 - 23: Four views of the MGS all nice and clean parked in an improvised motorpool in Iraq.

B-43

Pages 24 - 25: Various views of the MGS.

Page 25: Close-ups of the MGS mini-turret. The gun itself sits outside the vehicle on a rotating platform without much armor. The gunner and commander sit on either side of the gun.

ABOVE and LEFT: The commander's station with the .50 cal machinegun mount.

Page 27: The right upper side of the MGS, showing off the slat armor mounting brackets and the stowage bin.

ABOVE and LEFT: The gunner's station of the MGS shown in detail.

Page 29: The top of the M68A2 105mm gun turret.

COAX MG
ACCESS
CAUTION
NO STEP

ABOVE: The MGS wind sensor in its folded position.

LEFT: The commander's panoramic viewer is covered by a ballistic protective cover when not in the field.

ABOVE: A great detailed view of the commander's panoramic viewer.

RIGHT: The commander's station.

Page 32: The commander's station. Also visible is the layout of the slat armor brackets and armor itself on the vehicle.

ABOVE and RIGHT: The commander's station and the M2 .50 cal mount with shield.

ABOVE and LEFT: Two views of the right side and left side of the rear of the MGS gun turret.

ABOVE and RIGHT: Two views of the front and right front of the MGS. The above photo shows the driver's hatch open. Note the gun has to be traversed or elevated significantly in order for the driver to open his hatch.

ABOVE and LEFT: Two views of the front end of the MGS. The slat armor is simply just metal slats welded together to form a cage around the vehicle.

ABOVE: A good view of the Hull Protection Kit (HPK) side skirts.

RIGHT: The air conditioner system mounted on the MGS. Originally the MGS did not have any type of air conditioner system for the crew. They were issued cooled vests to wear inside the vehicle, but later models were fitted with an air conditioning system.

MV091S

MANEUVER CENTER OF EXCELLENCE
MV091S

Pages 38-39: A good walkaround view of the MGS at the Armor Training Center at Ft. Benning, Georgia.

RIGHT: Cutaway views of the 4 types of ammunition for the MGS. From Left to right in the photo: the M456A2 high explosive anti-tank round to destroy thin-skinned vehicles and provide anti-personnel fragmentation; M1040 canister shot for use against dismounted infantry in the open; the M393A3 high explosive plastic round to destroy bunkers, machine gun and sniper positions, and create openings in walls for infantry to access; and the M900 kinetic energy penetrator to destroy armored vehicles.

ABOVE: A front right view of the MGS showing the headlight assemblies and the other details.

LEFT: The commander's hatch open. You can see the handles of the M2 .50 cal machinegun also.

ABOVE and RIGHT: Two close-up views of the .50 cal mount on the commander's station.

Stryker MGS
In the Field

Saber Junction 2012

3/2 CR
K-42
3/2 CAV

I-42

3/2 CAV

Saber Junction 2015

MV0938
312 CAV
L-42

Wolfpack Supremacy 2012

038
PHONE
L-42

MV0938

MV0938

3/2 CAV
L-42

Grafenwoehr 2012

Dragoon Shara Khat 2010

2 SCR
G-42

2 SCR
G-42

4/2 CAV
AT-62

14
299